D0471456

The Book of Rhythms

Other books by Langston Hughes in the Opie Library

Popo and Fifina (with Arna Bontemps)

Black Misery

The Sweet and Sour Animal Book

The Book of Rhythms

Langston Hughes

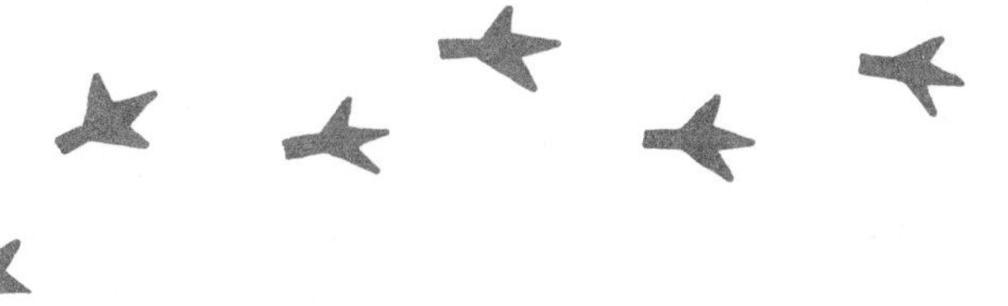

Illustrations by Matt Wawiorka

Introduction by Wynton Marsalis

Afterword by Robert G. O'Meally

Oxford University Press
New York

Oxford University Press

Oxford New York
Athens Auckland Bangkok Bombay
Calcutta Cape Town Dar es Salaam Delhi
Florence Hong Kong Istanbul Karachi
Kuala Lumpur Madras Madrid Melbourne
Mexico City Nairobi Paris Singapore
Taipei Tokyo Toronto
and associated companies in
Berlin Ibadan

First published by Franklin Watts, Inc. in 1954

Published by Oxford University Press, Inc.
198 Madison Avenue, New York, New York 10016

Design: Design Oasis

Library of Congress Cataloging-in-Publication Data
Hughes, Langston, 1902-1967
[First book of rhythms]
The book of rhythms / by Langston Hughes : illustrations by Matt Wawiorka : introduction by Wynton Marsalis : afterword by Robert G. O'Meally.
p. cm. — (The Iona and Peter Opie Library of Children's Literature)
Originally published as: The first book of rhythms. New York : F. Watts, 1954.
ISBN 0-19-509856-0
1. Rhythm. [1. Rhythm.] I. Wawiorka, Matthew, ill. II. Title. III. Series.
BH301.R5H8 1995
115—dc20 94-41270
CIP
AC

9 8 7 6 5 4 3 2
Printed in the United States of America
on acid-free paper

Contents

Introduction

Wynton Marsalis

Jazz musicians always say, "Rhythm is our business." Without rhythm, there is no music. I don't mean only jazz music, I mean any music. Rhythm is motion—no motion, no rhythm. No rhythm, no music. European classical music and church hymns and game songs and nighttime lullabies—all depend on rhythm to enliven their sound. The rhythm is the vitality of the music.

Take a simple song like "Happy Birthday." Sing its tune as usual. Okay, that's not bad. Now, clap your hands as you sing. Notice that a clapping rhythm gives the tune more vitality. It makes the tune happier, more joyous. Now try to sing "Happy Birthday" as if it had *no* rhythm at all—no predictable patterns that you could mark by clapping your hands. The song is ruined, right? The brightness dies.

Rhythm all by itself can be musical. Put your head down on a table and use your fingers to tap out the rhythm of "Happy Birthday"—*just the rhythm*. Check it out! Even with no tune and no harmony, the table taps can have the satisfying effect that only music provides. Likewise, in the hands of a master percussionist, a single drum can produce a wide range of musical colors and tones. It can dance, it can whisper, it can smile. As Langston Hughes said, "The tom-tom laughs, and the tom-tom cries."

In jazz music, all of the musicians have the responsibility of keeping the rhythm, not just the drummer. The piano and bass player are constantly drumming out rhythms on their instruments. That's why in jazz, the piano player, bassist, and drummer are called *the rhythm section*. Even horn players like me drum and dance the phrases out of our instruments. The fun comes when you take your listeners in one direction by playing some steady rhythm or repeated phrase, and then you change direction—jumping, skipping, or stopping unexpectedly. But this unexpected change is no mistake—it is calculated into the music. This tinkering with rhythm is called *syncopation*. Writers and visual artists do this, too. They put in a zigzag line or two or they change an angle or an attitude—they add a little pinch of spice that wasn't in the recipe but makes the result much more delicious.

When I was a small boy, growing up in Kenner, Louisiana, rhythms were all around me: the subtle rhythms of my father's piano, the rhythms of kids outside playing, and the rhythms of street parades with their dancers following the brass band. I played a few of those street parades, and believe me, they were something!

But in those days, no rhythm was more a part of the landscape of sound than the music of the trains that passed near our house. I'd hear the Southern Crescent and all the trains coming and going. I always remember that ka-nunk ka-nunk, ka-nunk ka-nunk, train-on-track sound and the woo-woo, woo-woo of its get-off-the-track whistle. When I play or compose now, I think about those rattling trains and try to get their sounds into my music. The trains have a romantic ring to them, too. At night I used to hear them in the darkness, coming from someplace mysterious, getting nearer, then going someplace far away.

That train sounded like a machine, but it also sounded like something human: a voice, a wail. It sounded like a lot of things and that's what made it fun to hear.

Jazz rhythm has that train sound. Even a real slow tune or a soft piece should have the intensity and swing as well as the mystery and romance of the train. The train is my symbol for rhythm in jazz music. If our band played "Happy Birthday," we'd want to make it swing just as hard as that rhythmic Crescent City train, bearing down like syncopated thunder.

As quiet as it's kept, I strive for these same qualities when I play Bach or Haydn, too. Even though classical music doesn't have the down-home dance-beat swing that jazz does, much of classical music *is* based on other kinds of dances which have their own rhythms. The more danceable the rhythm, the better the music sounds. Rhythm is the vitality of the music.

It pleases me that Langston Hughes wrote this fine book about rhythms because I grew up on Langston Hughes. When I was 9 or 10 my mother first read to me one of his short stories called "Father and Son," which is about pride and race violence. She also used to recite one of his poems that said, "Life for me ain't been no crystal stair." Even when his work is tough and hard edged, it fascinates me with its musical language and its rhythmic play. As a writer, he liked to swing his thoughts on the printed page. Rhythm was his business, too.

After you read this book you'll never see a seashell or a leaf or a train in the same way again—yeah, old Langston was hip to those trains, too!—yes he was.

Let's Make a Rhythm

It is fun to start something. Take a crayon or pencil and a sheet of paper and start a line upward. Let it go up into a curve, and you will have rhythm.

Then try a wavy line, and you will see how the line itself seems to move.

Rhythm comes from movement. The motion of your pencil makes *your* line. When you lift your pencil as you finish, the rhythm of your line on the paper will be the rhythm of your hand in motion. Try this:

There is no rhythm in the world without movement first.

Make a point of a triangle, then a smaller one, then a smaller one than that, then a still smaller one, so that they keep on across a sheet of paper, all joined together.

Again, you have made a rhythm. Your hand, your eye, and your pencil all moving together have made on the paper a rhythm that you can see with your eyes.

You can make a rhythm of sound by clapping your hands or tapping your foot.

You can make a body rhythm by swaying your body from side to side or by making circles in the air with your arms.

Now make a large circle on a paper. Inside your circle make another circle. Inside that one make another one. See how these circles almost seem to move, for you have left something of your own movement there, and your own feeling of place and of roundness. Your circles are not quite like the circles of anyone else in the world, because you are not like anyone else. Your handwriting has a rhythm that is entirely your own. No one writes like anyone else.

This is the way I write

How do you write?

Make a rhythm of peaks, starting from the bottom of one peak.

Make another rhythm like it, but start from the top.

Then do the same thing again, but put one rhythm over the other, and you have a pattern.

Fill in with your crayon and you have a pattern of diamond shapes. Rhythm makes patterns.

Draw a line. From the line make straight lines that slant up, each line growing a little longer. Or make another line and let the slanting lines cross it.

See how many rhythms you can make with straight lines, with curving lines, with circles overlapping circles, and with circles around circles, starting with a dot.

If a friend is drawing with you, see how different your friend's circles are from your circles. It is fun to make something yourself with your *own* rhythm because it will always be different from what anyone else will make. Your circles and rhythms are yours alone.

The Beginnings of Rhythm

Your rhythm on this earth began first with the beat of your heart. The heart makes the blood flow. Feel your heart. Then feel your wrist where your pulse is. That is where you can best feel the rhythm of your blood moving through your body from the heart. Doctors measure the force of the blood with an instrument which pictures this rhythm in terms of the speed of its movement. The speed varies from person to person, but when a line is used to picture this movement, it looks in general like this.

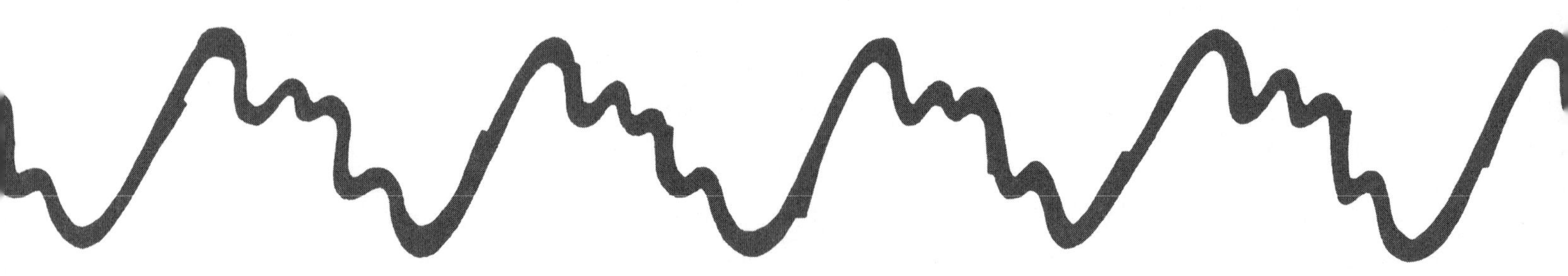

Such a diagram-drawing is called a graph. This particular line, you see, has a definite, repeating rhythm as the pulse throbs to the flow of blood pumped by the heart. The rhythm of life is the beat of the heart. The beat of the heart makes a pattern seen on paper when it is recorded by what doctors call an electrocardiograph machine. Sometimes the pattern looks like this graph.

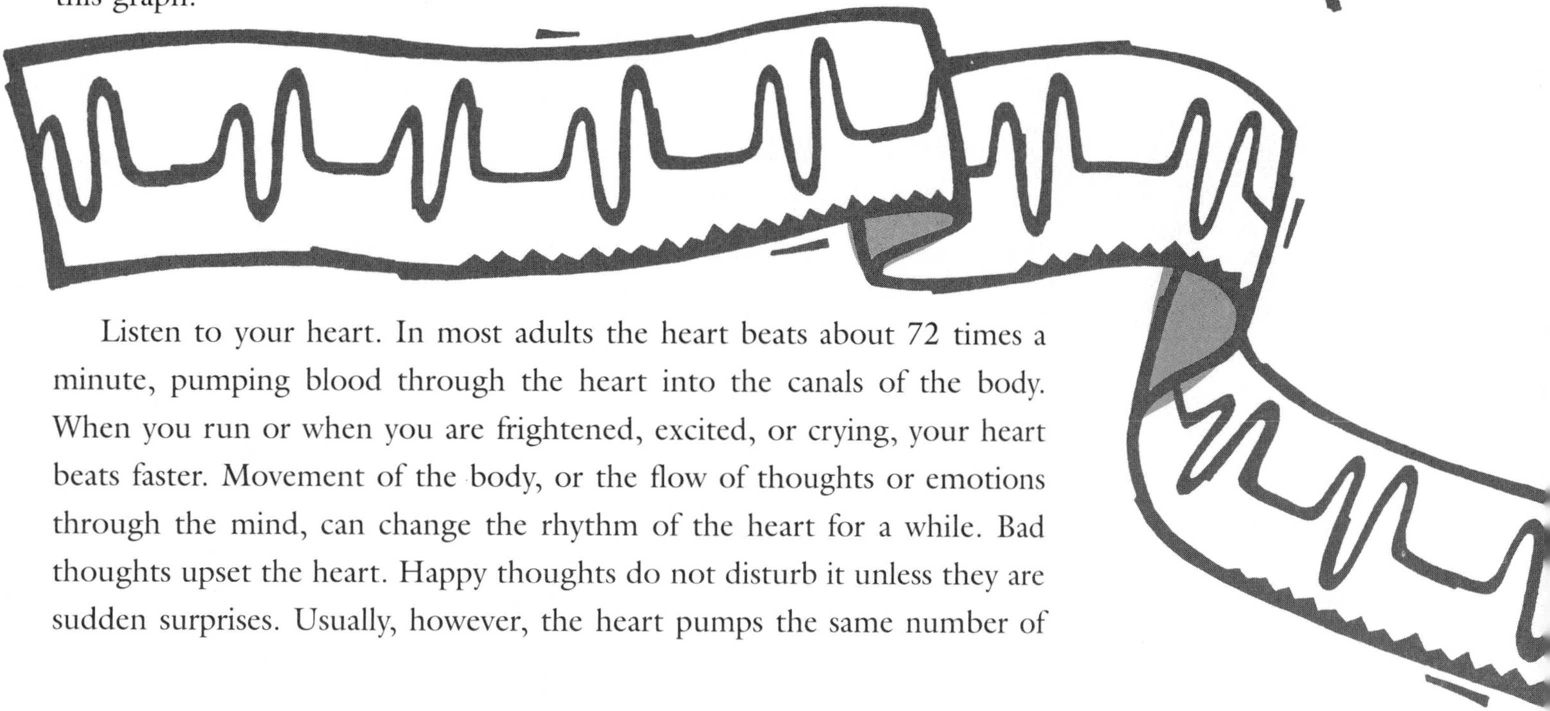

Listen to your heart. In most adults the heart beats about 72 times a minute, pumping blood through the heart into the canals of the body. When you run or when you are frightened, excited, or crying, your heart beats faster. Movement of the body, or the flow of thoughts or emotions through the mind, can change the rhythm of the heart for a while. Bad thoughts upset the heart. Happy thoughts do not disturb it unless they are sudden surprises. Usually, however, the heart pumps the same number of beats a minute, steadily, once a person becomes an adult, until he leaves our world. The rhythm of the heart is the first and most important rhythm of human life.

Thousands of years ago men transferred the rhythm of the heart-beat into a drumbeat, and the rhythm of music began. They made a slow steady drumbeat to walk to or march to, a faster beat to sing to, and a changing beat to dance to.

Try beating a slow steady rhythm with your fingers softly on the table, or on the edge of this book.

Try beating a changing rhythm that you can dance to. Try beating the rhythm of a song you know, like "The Star-Spangled Banner."

Try clapping your hands in rhythm as people do for square dancing, the Charleston, or for games.

The rhythms of music start folks to feeling those rhythms in their minds and in their bodies. That is why music sometimes starts the heart beating faster. One rhythm may start another. The rhythm of the wind in the sky will change the movements of a kite floating in the air. The rhythm of water in the sea will make a boat rock faster or slower as the water moves. Rhythm begins in movement.

A little stream starts to flow down a mountainside from a new spring. It makes a path across the land as the water moves. Gradually the path is washed out deeper and deeper, following the course of the water's movement. If the stream keeps on flowing for a long time it will make a gully. Over the years it may even create a canyon by the rhythmical flowing of the water. Out of a wilderness of rock the steady flow of the Colorado River carved America's beautiful Grand Canyon in Arizona, until it became an immense rhythmical cleft across the land. Its rhythm began with the steady movement of the water of the river. Even, steady motion is rhythm.

Varying Rhythms

The Mississippi River flows in a slow rhythm to the sea. Niagara Falls tumbles in a swift roaring curve of rhythm over a shelf of rock from one lake to another—Lake Erie to Lake Ontario. Water from an old-fashioned hose makes a single rhythmical arch as it falls onto the grass. But a garden spray produces a hundred lines of graceful curving water.

Steady, even rhythms are the easiest to make, or to look at, or listen to, or rest by. The tick of a clock, always the same, will put you to sleep. If you bounce a ball in an even rhythm, it is easier to keep it going. To skip rope and not miss, partners must turn the rope with an even speed. A swing swoops up and down evenly as you pump it. Grandma rocks steadily, not jerkily, in her rocking chair.

But steady, even rhythms are not always the most exciting, or the most interesting. An even rhythm is restful.

But an uneven rhythm is more interesting because it seems to be changing, to be going somewhere, to be doing something.

This rhythm does not change.

But this one does.

Varying rhythms are more exciting. This is perhaps why your heart beats faster when you start to school for the first time or take a trip or move to a new house and the rhythms of daily life change. Or why, when the rhythms of music vary or grow more rapid as in Ravel's "Bolero," or when the sea waves pound faster and louder, or the wind blows more swiftly and the trees sway violently, your heart beats faster, too.

One rhythm affects another rhythm. If one partner turns the end of a rope faster, the other partner must turn faster, too, to keep up. Otherwise there will be no rhythm in the turning. When two people dance together, they must dance together in the same rhythm, or it is no fun for either.

But different rhythms may sometimes be combined with interest. Straight lines make a rhythm of their own.

But with them a flowing line may be coupled to make an interesting combination of rhythms, one crossing the other, or one on top of the other, or below the other.

There are many combinations of varying rhythms and actions. A merry-go-round goes round and round while the animals on it go up and down. If you have ever watched a sea gull flying you have noticed that it flaps its wings awhile, then glides awhile, then flaps awhile, then glides again. Flap, flap—glide—flap, flap—glide. See what other kinds of varying rhythms you can think of yourself.

Sources of Rhythm

Artists have used animals, trees, men, waves, flowers, and many other objects in nature for rhythms. In France 25,000 years ago the cave men made animal drawings on the walls of their caves.

Later the flag lily, *fleur-de-lis*, became a rhythmical design that is the national symbol of France.

African artists a thousand years ago made beautiful masks with rhythmical lines.

In the sixteenth century a Spanish artist named El Greco sometimes made a man look like this.

And Picasso with a few lines has made a bull like this.

An American artist named Paul Manship made a figure of a man in Rockefeller Plaza in New York like this.

Each artist makes his own rhythms out of the things he sees around him and sees in his dreams and mind.

The most beautiful rhythms seem always to be moving upward. That must be because the sun is above, and the growing things that start in the earth grow upward toward it.

A blade of grass moves upward as it grows. So does a flower, first only a little sprout, then a stem giving off other stems that turn into branches that bud as the plant grows taller. The buds turn into flowers as high above the earth and as near the sun as each plant can reach. Small plants like violets do not reach very high. Tall plants like sunflowers or hollyhocks sometimes grow taller than a man. Trees like magnolias bloom away up in the air. From the roots in the earth to the tallest flower, lines of rhythm flow upward. And in each leaf, each flower petal, there is rhythm. Nature is rhythm.

The lines of an oak leaf move outward and upward to its tip.

A locust leaf has rhythm, and a violet in bloom.

There is rhythm in a lily, and the bud of a rose before it opens much, and a many-petaled rose in full bloom.

Many vines curl upward. But some plants, like Spanish moss which lives on air, hang gracefully downward from the branches of trees. The leaves of water lilies float calmly on the surface of the water, but their blooms open toward the sun.

In growing things there is an endless variety of rhythm from the shaggy, pyramiding pine to the tall bare curve of a towering palm, from the trailing weeping willow to the organ cactus of the desert, from the straight loveliness of bamboo trees in the tropics to the wind-shaped cypress of the California coast, clinging to a rocky cliff near the sea where the waves shower their salty spray.

This is one of the many rhythms of moving water in the sea.

These are some of the many thousands of rhythms that moisture makes when it forms a tiny snowflake, or when it becomes a ball of hail, or an icicle, or when it is just, as many poets have said, "the harp-strings of the rain."

Perhaps falling rain first gave men the idea of painting stripes as decoration on their walls or on their war shields, or of weaving stripes into their garments, or making a rhythm of slanting stripes that lean as rain sometimes does, or stripes flowing as rain in the wind may seem to flow and curve.

Perhaps the curve of a waterfall or the arching stripes of the rainbow suggested the rhythms for the arches of the houses and temples and tombs and bridges of men long ago—the arch of Tamerlane's tomb at Samarkand, the arch of a bridge in ancient China, or the Moorish arches at Granada.

When the Egyptians built their tombs and temples over a thousand years before Christ, they knew how to combine the rhythms of nature with the possibilities of stone and sun-dried brick in the structure of their buildings. And the more harmoniously they did this, the more

beautiful were their buildings. In splendid palaces the pharaohs lived.

The Greeks, hundreds of years before Christ, knew the rhythmical beauty of the soaring line in a column. The rising lines of its many columns made the Parthenon one of the most beautiful buildings ever created.

The columns of Greek temples go upward.

The pyramids of the Egyptians point upward.

The skyscrapers of American cities rise into the skies.

Like the blades of grass and the stems of flowers and the trunks of trees, the houses and temples and other buildings of man rise toward the sky where the sun is. Almost nobody builds a house, church, or any kind of building underground.

The Rhythms of Nature

From the motion of the planets around the sun—Old Sol—the rhythms of the sky are charted. The sun and the planets which move around it, with the moons, the comets, the meteors, and all the smaller asteroids too far away to be seen, are called the solar system.

You with your own eyes can watch the daily rhythm of night, day, night, day, night, day, and the rhythm of the four seasons that are spring, summer, fall, and winter, regularly repeating themselves over and over again. The rhythms of the moon becoming full, then waning, then becoming full again, you can see, too, as it revolves around the earth. The magnetic pull of these rhythms is recorded in the tides of the sea.

In nature many rhythms are related, one to another, and men and animals live according to these rhythms. Fields are planted in the spring, cultivated in summer, harvested in the fall, and the grain is stored for winter

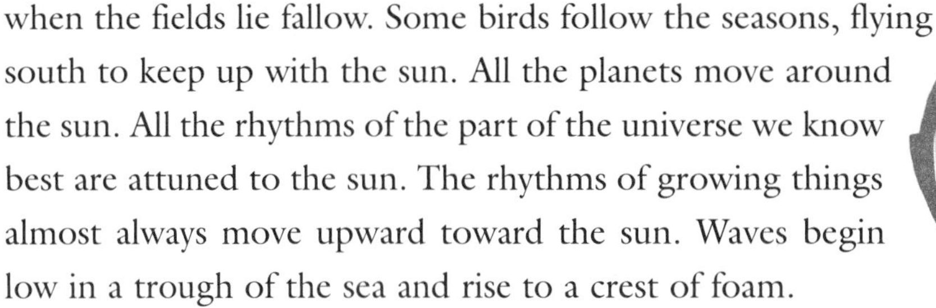

when the fields lie fallow. Some birds follow the seasons, flying south to keep up with the sun. All the planets move around the sun. All the rhythms of the part of the universe we know best are attuned to the sun. The rhythms of growing things almost always move upward toward the sun. Waves begin low in a trough of the sea and rise to a crest of foam.

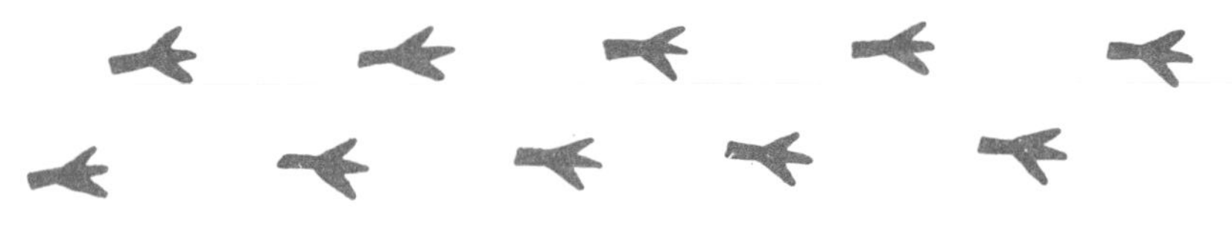

The rhythms of the joyful spirit are rising ones. Animals in happiness leap upward. Ballet dancers rise as high as they can on their toes when dancing. Church steeples point toward the sky. When men pray their thoughts go up, and their souls are uplifted. When people are happy they generally walk with their heads up. When they are sad and days are dark and sunless, their heads are often down.

The sun influences the moon. The sun and moon influence the sea. The tides of the sea influence the shell life of the sea, and each shell is molded into a rhythmic shape of its own by all the rhythms and pressures that bear upon it from the sea. These are a few of the many shell shapes.

On the surface of each shell the lines of some additional rhythms are etched in graceful beauty.

No one shell in the world is exactly like any other shell. There are millions of different shells. Even each shell of each family is different and the lines on each single shell are different, too.

You, like the shells, are like no one else. Even the rhythm of each line in your hands is like that in no one else's hands.

The rhythm of your walk is like no one else's walk. Some people walk with long easy steps, some trip along, some wobble. Soldiers learn to march in step together, with their walking rhythms at the same tempo, so that they all cover the same amount of ground in the same time at the same speed. They look better marching in rhythm. If one man gets out of step, he makes the whole company look bad. That is why all must follow the same marching rhythm. Forward, march!... One-two-three-four!... One-two-three-four!

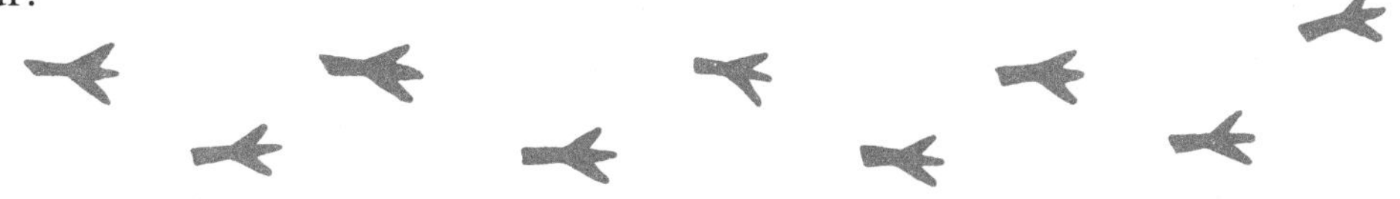

At the Radio City Music Hall in New York there is a famous chorus of girl dancers called the Rockettes. Each girl's foot strikes the stage at the same time in rhythm when they are dancing. They are called precision dancers. When people are dancing together, the nearer their rhythms are alike, the better they look.

Rhythms of Music

Music sets the rhythm for dancing. The body follows the beat of the music. There are the fast whirl of the Viennese waltz, the skip and hop of the square dance, the glide of the fox trot, the step-step of the two-step, the fling of the Charleston with its flying legs, the prance of the old-fashioned cakewalk—each having its own style and rhythm. The first dances were done to the rhythms of the drums, long before other instruments were invented. And the tone of the drums and the volume of their sounds were molded by the rhythm of their shapes.

In America some of the happiest music for dancing is made by our jazz bands whose rhythms were born from the drums of Africa, and changed over to the piano, the trumpet, the bass viol, the saxophones, and all the other instruments that make up a band.

There is rhythm for dancing in words, too, like "Waltz me around again, Willie," the title of a song popular fifty years ago, which has a dancing sound.

The rhythms of music and of words cause people to want to move, or be moved, in time to them. Babies like to be bounced to

Bye, Baby Bunting,
Papa's gone a-hunting
To get a little rabbit skin
To wrap the Baby Bunting in.

Children dance to

Ring-around-the-rosy,
Pocket full of posy!

or play a game to

London Bridge is falling down,
Falling down, falling down!
London Bridge is falling down,
My fair lady!

In music there are simple rhythms and complicated rhythms. A popular song may follow the same rhythmical beat all the way through. In a symphony there may be many varied rhythms. The conductor of a large orchestra sets the tempo and controls the rhythms with the movement of his baton. If the men do not follow him the music is badly played and not pleasing. In the barnyard when animals are frightened and all the chickens start cackling, the dogs barking, ducks quacking, and cows mooing, each in different rhythms, the best way to describe the noise is by the word "pandemonium," which means a disorderly, inharmonious uproar. No one is moved by such noises.

But men are often moved by rhythmical music to work together better. Sailors on the old sailing ships sang sea chanteys as they hauled in the sails or lifted the anchor:

Heave-ho! Blow the man down!
Oh, give me some time to
Blow the man down!

Men plowing or sawing or breaking rock like to work to music, with the rhythm of the song matching the rhythm of their bodies as they work. In this song from the deep South the "Huh!" is where the men's breath comes out as their hammers hit the rock or their picks strike the earth:

> I got a rainbow... Huh!
> Round my shoulder... Huh!
> Ain't gonna rain!... Huh!
> Ain't gonna rain!... Huh!
> This old hammer... Huh!
> Killed John Henry... Huh!
> Won't kill me, Lord!... Huh!
> Won't kill me!

Rhythm makes it easier to use energy. Even sweeping a rug or raking a lawn is done better if the broom or rake is handled with a steady rhythmical motion. In countries where a farmer still harvests the grain by hand, he advances across the field with the rhythmical sweep of his scythe, and the grain falls before him. Often he sings as he mows. And at the end of the harvest all the farmers sing for fun.

Rhythm and Words

Even without music there is rhythm in verses like these:

Sing a song of sixpence
Pocket full of rye!
Four-and-twenty blackbirds
Baked into a pie.
When the pie was opened
The birds began to sing.
Wasn't that a dainty dish
To set before a king?

Rhythm is very much a part of poetry. Maybe that is because the first poems were songs. In ancient Greece poets made up words and tunes at the same time. In the Middle Ages bards and troubadours sang their poems.

Nowadays poets usually make up only words. But behind the words of good poems there is always rhythm. These lines by William Blake have a fine rhythm:

> Tiger! Tiger! Burning bright!
> In the forest of the night!

Even when poems do not rhyme there is rhythm, as in the beautiful poetry of the Bible:

> Whither thou goest, I will go,
> And where thou lodgest, I will lodge.
> Thy people shall be my people,
> And thy God my God.

Or listen to the rolling lines of Walt Whitman:

> In the dooryard fronting an old farmhouse near the white washed palings,
> Stands the lilac-bush tall-growing with heart-shaped leaves of rich green,
> With many a pointed blossom rising delicate, with the perfume strong I love,
> With every leaf a miracle....

Stories and sermons and speeches and prayers have their rhythms, too. How beautiful is the rhythm of Lincoln's *Gettysburg Address:*

> Fourscore and seven years ago our fathers brought forth on this continent a new nation, conceived in liberty, and dedicated to the proposition that all men are created equal.

Just as the feet of soldiers marching in rhythm carry men forward, so the rhythms of sermons or speeches or poems carry words marching into your mind in a way that helps you to remember them. That is why I think

> To make words sing
> Is a wonderful thing—
> Because in a song
> Words last so long.

Most of the rhythms men put into music and poetry may be found in nature: the drumming of the rain, the rap-rap-rapping of a woodpecker on a tree, the steady beat of the waves on the beach—breaking, retreating, breaking, retreating over and over again. The call of the whippoorwill, the bobolink, the oriole, the sparrow, the trilling song of the canary—all these helped men to form their own rhythms and make their first music.

Some Mysteries of Rhythm

Deep inside of men and animals there are other rhythms that we cannot explain, but that are a part of life. Nobody knows why, for example, different kinds of birds always build their own particular kind of nests, each with its own peculiar rhythmical shape.

A South American ovenbird's nest is always oven-shaped.

A Baltimore oriole's nest hangs down like a hammock.

Every worker honeybee that lives knows how to help build a honeycomb, and the cells of each comb are always the same shape and of the same two sizes.

White-faced hornets' nests are always like this.

How each new brood of birds or bees or hornets knows how to create the same patterns and the same rhythmical shapes we do not know, but we call it instinct. This is nature, handing down certain ways of living and

rhythms of life from one generation to another. Old swallows and young swallows all swoop through the skies in the same arcs of motion, and wild geese always fly in the same V-formations. For some reason wild geese find it easier to fly in formation this way.

Men and women create rhythms that make things easier to do, too. In fact, some things could not get done at all without rhythm. If you live in the country, watch a farmer churning butter. If he does not churn with an even up-and-down rhythm there will be no butter. Watch your mother beating cake batter. If she beats this way, that way, then slow, then stopping, then fast, the cake will not come out a good one. But if she beats her batter with a steady rhythm everything is mixed well and it is a good cake.

Athletics

When you go to the circus and see a man or a woman in spangled tights swinging on a trapeze high at the top of the tent, you remember for a long time the thrill as the trapeze artist swoops gracefully into space. The rhythm of a swimmer leaping from a diving board, cutting through the air in the swift curve of a high dive, is a beautiful picture to remember, too. In the water, the swimmer with the smoothest and most harmonious rhythm to his strokes is the one who moves the fastest and most beautifully through the water.

The word "harmonious" is a part of the definitions of rhythm in most dictionaries. Rhythm is "an harmonious flow," says one. Another dictionary says that rhythm is the measure of time or movement by regular beats coming over and over again in harmonious relationship. A good athlete must have that harmony of movement or rhythm, which is called "form."

Notice the form of a championship baseball player. Pitchers like Allie Reynolds or Satchel Paige or Ewell Blackwell wind up for the pitch and let

go with a rhythm that begins in the tips of their toes and ends at the tips of their fingers as it is transferred to the ball that speeds through the air.

As for batters, a whirlwind is like this:

A seashell is like this:

Stan Musial batting a ball is like this:

Watch Johnny Mize or Ted Williams, Ralph Kiner—or yourself—for when you bat a ball, your whole body turns with the bat. A rhythm is set in motion that, in its turn, carries the ball, if you hit it well, in a curve through space. The ball makes its own rhythm in the air as it moves into the outfield. The outfielder leaps into the air like a bird as his hands go up to catch the ball. From pitch, to swing, to ball, a whole series of rhythms are set off, one rhythm, or one motion, starting another. So it is in life—from sun, to moon, to earth, to night, to day, to *you* getting up in the morning and going out to play a game of ball. *All* the rhythms of life in some way are related, one to another. You, your baseball, and the universe are brothers through rhythms.

Broken Rhythms

When rhythms between people are jerky and broken—for example, when friends quarrel—life is not happy. When you row a boat you cannot make much headway with short choppy strokes. But when your body bends to the strokes of the oars with a slow easy rhythm you can feel the transfer of your rhythm to the oars. Then the boat, too, takes on a rhythm and moves forward swiftly and smoothly through the water. People and races and nations get along better when they "row together."

When a sail is unfurled properly on a sailing ship and billows out as it catches the wind you can sense its rhythm and almost hear the music of its motion. You can almost see the wind in the canvas. When a kite flies you can feel at the string's end the way it is pulled by the wind. When you spin a top you can feel in your throwing of the top the way your rhythmical unfurling of the string changes into a hum and a whirl. When ballet dancers spin on their toes they spin in rhythm. If they do not, the results are ugly. Broken rhythms usually are not beautiful.

In 1910, the lines of automobiles were not as beautiful as the smoothly flowing lines of today's models, but were broken and full of angles. And even if they had had better motors these older automobiles with their clumsy, angular, broken rhythms could not have moved as fast as the new cars with their flowing lines, for the rhythms of shape are related to the rhythms of movement.

Machines

But new things are not always better than old things. The lines of a Grecian water jar made three thousand years ago are more beautiful than some modern water jugs. One reason is that in the old days more things were made by hand. Therefore many things were different each from the other because each man, when making something, put into it his own individual rhythms.

Nowadays many things are made by machines that turn out thousands of copies, all just alike. When some of the long-ago Indians wove cloth by hand, each cloth was different and usually beautiful. Machines turn out beautiful cloth, too, and more of it, faster, by making over and over the same copies. They may be beautiful copies *if the first designs are beautiful.* But a man or a woman must *first* design the cloth. Machines cannot create beauty. They can only copy it.

Machines move in rhythm. You can hear the rhythm of a printing press. You can see the rhythms of a generator in motion making energy to start, in turn, the motion of the machine it serves. You can see the rhythms of a piston, of a ramrod, of a shuttle, of a washing machine, of an egg beater, of a propeller.

An airplane propeller thrusts forward into currents of air and on the rhythms of the wind a plane rises into the sky. A Diesel engine turns its rhythms into power that carries a streamlined train rocking across the rails from city to distant city. The rhythm of a clock's balance wheel can be heard in its steady tick-tock, tick-tock, tick-tock. The rhythm of a steam shovel can be seen as it scoops up tons of earth.

Rhythms May Be Felt —and Smelled

The rhythms of the wind can be felt and certain sounds can be both felt and heard. An explosion at a distance can knock you down. Certain tones on the piano or violin hurt a dog's ears and will set him to howling. Some electronic vibrations, like those from short-wave radio transmitters, men cannot hear, but a dog can hear them, or feel them in his ears, and they hurt his eardrums. The vibrations from the sharp tone of a flute in a room have been known to break a pane of glass in a window. The flute player does not feel this tone from his flute, but it hits the pane of glass as hard as a stone, and the glass feels it. You cannot see electricity, but a shock can knock you down. Such an unseen rhythm may take on the force of a moving body. Perhaps tomorrow men may learn how to use such rhythms to shoot space ships to the moon.

In chemistry there are tiny bits of matter called molecules, too small for the eye to see. They move in a definite rhythm in liquids or gases, but we cannot see each tiny particle that makes up the gas. The air itself is gas, but we cannot see it. Instead we see *through* it.

Odors, or scents like those of perfumes, are really tiny particles, gases, or vapors in rhythmic motion. We smell them when they touch our noses although we cannot see them. In both chemistry and physics there are rhythms so complicated that it takes years of study to understand them. The rhythms of the patterns of light may be seen in colors called the spectrum. But there are many rhythms we cannot see, but which we can chart or measure.

Unseen Rhythms

Music may be heard, but not seen. It is true that the notes of music may be seen as written down on paper. But notes serve only as a guide to sound. They mean very little until they are set in motion by the human voice or by instruments, to travel in rhythmical pulsations—like the beat of the pulse—to the ears of listeners. Sir Isaac Newton, who was a great pioneer in the science of modern physics, called sound "pulses of air." Sound cannot travel far alone, but from a radio tower, through the power of electronics, sound waves are transformed into electrical pulsations which go out on invisible electro-magnetic waves in every direction. Then people thousands of miles away can hear the sound by means of electronics. Our President can make a speech in Washington and folks across the ocean in London can hear him because of this marvelous modern science.

Electronics is the science of the movement of electro-magnetic waves. In broadcasting, these waves move outward and upward and around the earth through the air. They silently carry with them sound waves in electric pulsations which become audible words or music by means of our radio sets. They carry sights, too—unseen as they travel through space—which become pictures for our television screens. They create invisible radar guides which may direct an airplane through a fog to land, or keep a ship away from a reef.

There is an instrument called a phonodeik (FO-no-dike) for picturing the vibrations of sound waves as they travel.

This is the rhythm of a single tone in a soprano voice as it moves through the air.

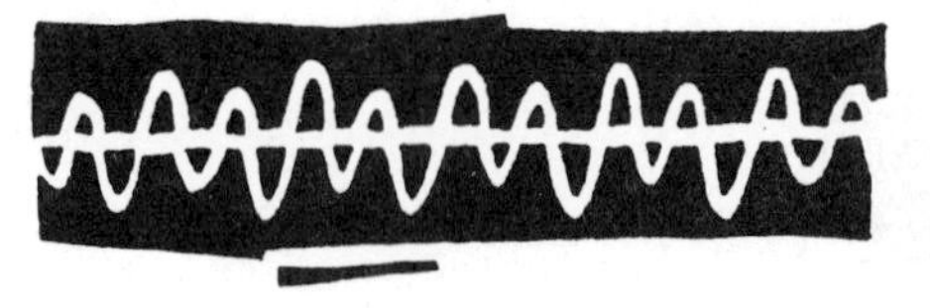

This is a graph of the tone of a violin.

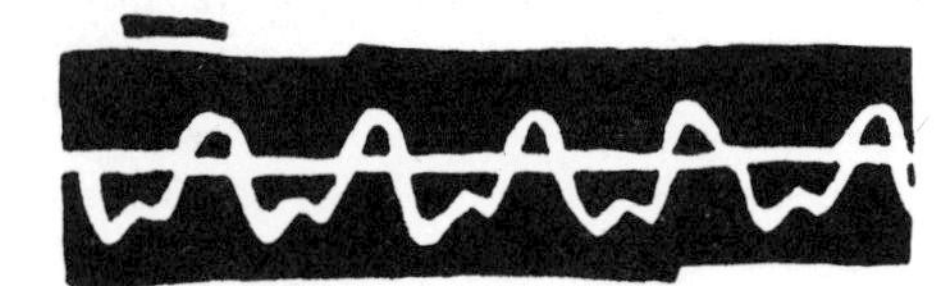

This is a bell tone.

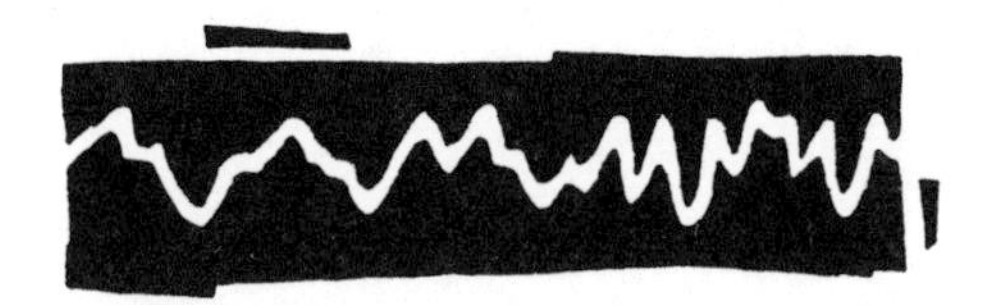

And in the field of electronics, if the radar waves moving through space *could* be seen, they would look like this.

The rhythms of electronics were a mystery to men a hundred years ago. But now we know that light and sound and the force of the atom, too, each travels in its own series of rhythms. So, in this wonderful universe of ours there are rhythms that men can see, like those of sea waves. And there are unseen rhythms that we cannot see, but which we can chart and measure by scientific instruments and by mathematics. There remain to be solved tomorrow many fascinating mysteries about these unseen forces.

Rhythms in Daily Life

What is exciting about tomorrow is that it is always different from today. Back in the early 1900s the first airplane looked like a crate. But modern airplanes are streamlined.

To streamline something means to make the lines more rhythmical, more like a stream flowing. It is much more pleasing to look at a rhythmical, harmoniously shaped object than at an awkward thing. That is one reason industrial designers who create many of the things we use in daily life are paying more attention to rhythm and beauty.

Old-fashioned trains looked this way, and the coaches were rectangular.

New trains look like this.

The old ice box was shaped just like a box. But today's refrigerators are more graceful. And they are chilled by electricity from a dynamo whose power-rhythms are generated many miles away from your kitchen.

In grandma's day in the country, butter was put into a well to keep fresh. And cool water was drawn from the well in an "old oaken bucket." Thousands of years ago, before men learned to dig wells, they drank from springs or streams.

The first bowl to hold water was a man's cupped hand, or a pair of cupped hands.

Then the hands took clay and made a ball. A bowl is half of a ball, hollowed out. And a cup is just a bowl with a handle on it. A plate is just a bowl flattened out.

The first bowls and cups and plates were made only to be useful. Now we make them to be beautiful as well. Today fine designers like Russel Wright make beautiful dishes that are reproduced by machine methods so that people may buy them cheap. The rhythms of the hands of many makers of bowls, cups, glasses, and plates have given us the thousands of beautiful and useful shapes from which we eat and drink.

Furniture

It is interesting to observe the rhythms of furniture. The first chair was probably just a rock or a fallen log. Then a part of a log was set upright with a piece of the trunk left on it for a back. Then some planks were put together in angular fashion, very straight up and down.

Now chairs are made in many charming and graceful rhythms. Harmony or rhythm is very much a part of modern furniture-making.

And in our wallpaper and cloth, our towels and our curtains, our scarfs and neckties, the rhythms of lines and design are many. From the rhythms of nature—the pistil of a flower, the shape of an acorn, the tips of wheat stems—may come the designs on a man's tie, the patterns on living-room drapes, the border around your bedroom. At home, look and see what inspired the shapes and designs in your mother's wallpaper or the drapes at the windows.

How Rhythms Take Shape

Rhythms always follow certain conditions. Knitting creates its own rhythmical patterns by the very way the needles work in wool. So does chain stitching, cross-stitching, or featherstitching in sewing. Threads in weaving follow a rhythmical pattern according to the kind of weave being made. Pleats in dressmaking—knife pleats, accordion pleats—have a rhythm growing out of the way they are folded. When mothers curl little girls' hair they simply put the hair into the rhythm of spirals. But curls do not look well on all girls. Each person should arrange her hair to suit the shape of her face, just as a well-dressed woman chooses her hats to suit the lines of her profile, or her gowns to go well with the lines of her body.

Fashions in clothes and styles of decoration and design are everywhere influenced by other people and places around the world. That design on your father's tie may be from Persia. Your uncle's jacket is a Scotch plaid. The lines of your mother's dress are French. Her shawl is Spanish in design.

The music on your radio now is Cuban, its drums are the bongos of Africa, but the orchestra playing it is American. Rhythms go around the world, adopted and molded by other countries, mixing with other rhythms, and creating new rhythms as they travel.

In the Arctic the igloos make a curved rhythm against the sky. In America skyscrapers rear geometric towers above us. In Bechuanaland there are beehive huts, in Japan slant-roofed pagodas. Each building has its rhythms, loved by the people who made them, and often borrowed by others elsewhere. The Quonset hut of the American army is in shape not unlike an Eskimo igloo. And the East Indians are building skyscrapers in Bombay. We love Irish folk songs. The Irish love American jazz. Sometimes American jazz and Irish melodies combine to make a single song.

This Wonderful World

How wonderful are the rhythms of the world! Poets everywhere write about the drowsy hum of a bee. Farmers everywhere wake up to the cascade of a rooster's crow. In nature the rhythms we hear range from the trill of a bird to the chirp of a cricket, the lonely howl of a coyote to the thunderous roar of a lion, the gurgle of a brook to the boom of the sea, the purr of a cat to the beating of your own heart.

Look at the upward sweep of the horns of the antelope! See the outward curves of the Texas long-horned steer, the antlers of the elk, the stripes of the zebra or the tiger, the graceful neck of the giraffe, the delicate hoofs of a goat on a ledge, the curve of a sea lion on a rock, the scoop of a cat on a pillow, the flicker of a fish, the leap of a monkey, the wiggle of a puppy, the dive of a heron, the balance of hummingbirds, and butterflies, and ballet dancers, the drift of clouds

across the sky in moving masses of vapor, the ever-spreading rhythm of the circles when you throw a stone into a pool of still water, the unseen rhythms of electronics that come right into your house bringing your favorite programs onto your TV screen! When a radio tower projects its electronic waves into the air, someone in Paris can hear a voice in New York saying, "Good morning"! Yet you cannot see or hear that voice on its way across the ocean. Such are the seen and the unseen, the heard and the unheard rhythms of our world.

The earth which is our home moves in its own rhythm around the sun, as do all the planets. Animals, and boys and girls, and men and women, get up in the morning by the sun. Plants live by the sun. The moon moves around the earth as well as the sun. The rhythms of the sun and the moon influence the sea, the seasons, and us, and

affect the rhythms of our universe, so immense in time and space that men do not yet understand most of it.

But your hand controls the rhythms of the lines *you* make with your pencil on a paper. And your hand is related to the rhythms of the earth as it moves around the sun, and to the moon as the moon moves around the earth, and to the stars as they move in the great sky—just as all men's lives, and every living thing, are related to those vaster rhythms of time and space and wonder beyond the reach of eye or mind.

Rhythm is something we share in common, you and I, with all the plants and animals and people in the world, and with the stars and moon and sun, and all the whole vast wonderful universe beyond this wonderful earth which is our home.

Afterword

Robert G. O'Meally

From the mid-1920s until his death in 1967, Langston Hughes was a key figure in American letters. Richly prolific, Hughes wrote poems, novels, short stories, plays, essays, song lyrics, and works for children. His writings have been translated into many languages; some are among those magic American works, like *Huckleberry Finn* and *The Call of the Wild,* that are read not only as homework assignments but are picked up by young readers for sheer pleasure. In libraries, his stories of Jesse B. Simple (the wry, unsinkable black Everyman whose name gives a hint of his philosophy for survival) and his *Selected Poems* are thumbed to pieces by eager readers of every description. For 70 years Hughes's works have been recited at church and school programs, invoked at graduation ceremonies, and read as bedtime pieces all over America and beyond.

Throughout his career, Hughes wrote literature that appealed to children. He delighted in his young audience. Not only did it promise good readers for the future, but this juvenile group was valued for its own sake. Kids responded to his poetry, and he loved them for it. Hughes believed that young people benefited greatly by writing poetry of their own. In an unpublished essay called "Children and Poetry" (1946), he said that he detected a "psycho-therapeutic value in that, unconsciously and by indirection, they may get down on paper some of the things that trouble them—and thus relieved, live better, freer, less confused lives."

On the jacket of the first edition of *The Book of Rhythms* (originally published as *The First Book of Rhythms* in 1954), Hughes was introduced as a famous poet whose present book was inspired by his teaching experiences at the Laboratory School in Chicago. The Lab School, an experimental school for children from kindergarten through 12th grade, was defined by John Dewey's ideal of "learning by doing." In 1949, Hughes spent a term teaching an interdisciplinary course for eighth graders called the Special Arts Project. According to his carefully wrought lesson plans, Hughes led a series of discussions of rhythm in plants, animals, and the universe, as well as rhythm in human body movement, speech, music, visual arts, and, of course, poetry. His notes indicate that for homework students were to prepare reports comparing rhythms they had observed. One was scheduled to make a presentation on music and baseball, another on swimming and modern dance.

In 1951, when Hughes addressed the annual meeting of the College Language Association, he drew on his months at the Lab School to recommend "Ten Ways to Use Poetry in Teaching." That lecture's definition of poetry again forecast the present volume's most pronounced themes. He said: "Poetry is rhythm—and, through rhythm, has its roots deep in the nature of the universe; the rhythms of the stars, the rhythm of the earth moving around the sun, of day, night, of the seasons, of the sowing and harvest, of fecundity and birth. The rhythms of poetry give continuity and pattern to words, to thoughts, strengthening them, adding the qualities of permanence, and relating the written word to the vast rhythm of life."

The Book of Rhythms is from one perspective an extension of Hughes's teaching efforts and part of an ongoing project of addressing young readers with material that was playful but broadly meaningful. But it also represents part of Hughes's effort to make and sell art in the lean and politically repressive 1950s. Though his income was solid, his expenses were onerous. As he wrote in mournful tones to his friend Arna Bontemps in December 1953: "Another year of starvation. My art costs me more than I make.... Precious Lord, take my hand!"

That year, while *The Book of Rhythms* was in progress, Hughes—whose writing in the 1930s had been sharply leftist in its orientation and whose writing throughout his career had advocated freedom for black Americans and oppressed people everywhere—was summoned before Senator Joseph R. McCarthy's notorious congressional committee investigating those suspected of supporting communism and other "un-American activities." In difficult private and public sessions, Hughes responded with anguish to the committee's barbed questions about his forthrightly politicized verse, some of it written 20 years before the hearings. Maintaining a degree of composure and dignity, he defended himself as a loyal American who was sometimes critical of his nation. But he was not a communist, he quietly said, and no longer wished to be judged by his socialist poetry of the 1930s.

To support his case, Hughes recommended that the senator look at some of his recent children's books, including *The First Book of Negroes* (1952), where he had written that in spite of "many problems," America was a place where "people are free to vote and work out their problems. In some countries people are governed by rulers, and ordinary folks can't do a thing about it. But here all of us are a part of democracy. By taking an interest in our government, and treating our neighbors as we would like to be treated, each one of us can help make our country the most wonderful country in the world." (Of course, these words were put to paper, as Hughes knew as well as anyone, at a time when blacks in the South were systematically excluded from voting, and where even the most muted forms of political assertion by blacks could result in very violent responses.)

Released, through compromise, understatement, and overstatement, from the grasp of the McCarthy committee, Hughes nonetheless continued to suffer other effects of the cold war and its literary witch hunts, book purges, and blacklists. Hughes had begun to write children's titles that no librarian could question (he reported to friends that his books were getting "simpler and simpler and younger and younger"). But it is ironic that in working with children's publishers Hughes found himself trying to please some of the most politically timid people in the business. For example, in light of Hughes's reputation as a leftist, Helen and Franklin Watts—the publishers of *The First Book of Negroes*—required him to submit a statement assuring their readers that he was not a communist.

To be on the safe side, *The First Book of Negroes* featured no mention at all of Richard Wright, Paul Robeson, or of Hughes's hero since boyhood, W. E. B. Du Bois: all were too closely identified with the Left. Even Josephine Baker, pictured in the first edition, was cut out of the book when a New

York columnist threatened to pan it because he believed Baker to be a communist. Hughes's *Famous American Negroes* (1953), also devoid of political leftists like Du Bois, was heavily edited by Dodd, Mead's editors, who removed all of Hughes's references to overt racism. For example, the editors cut 19 lines from the sketch about the entrepreneur Charles C. Spaulding, who was, Hughes had reported, an elderly man struck in the face by a white drugstore clerk working in a building owned by Spaulding's bank because the clerk had seen Spaulding sipping a beverage while on the premises.

Of course, the 1950s represented a period not only of chill and caution but of the promise of dramatic political change. Doubtless, Ralph Ellison's *Invisible Man* (1952) offers the most telling metaphors for that complex season of distress and anticipation: Ellison's black would-be hero is underground and invisible but all the same he senses a world of "infinite possibilities" just around history's corner.

On May 17, 1954—at about the time *The Book of Rhythms* arrived in bookstores and libraries, the Supreme Court, in *Brown* v. *Board of Education,* ruled 9 to 0 that school segregation was unconstitutional because "separate education facilities are inherently unequal." Nineteen fifty-five was the year of Rosa Parks and the Montgomery bus boycott, which led to Martin Luther King, Jr.'s ascendancy to the leadership of the civil rights movement. These monumental events defined that era at least as much as the repressive energies of the cold war did.

It is also true that landmarks beyond the world of politics and letters added to the period's contour and significance. In 1951, Sugar Ray Robinson helped inscribe the decade as one of black heroic achievement by taking over as middleweight boxing champion of the world; in baseball, Willie Mays won the National League home-run championship in 1955 (and then again in 1962, 1964, and 1965). It was in 1955 that Marian Anderson became the first African American to sing at the Metropolitan Opera House in New York City.

Rock-and-roll music, whose earthy black-and-blue rhythms were exploding across the tracks to middle-class white America (and everywhere else), offered another idea of the 1950s sound. But perhaps the highest honors here should go to jazz players like Miles Davis and his band. Davis's muted trumpet—sometimes coolly understated and plaintive in its lyricism, and sometimes caustic and fiery with impatience, a blue-silver jet of barely controlled anger—told its tales above a rhythm section famous for its tightness and crackling energy. Davis's horn seemed to declare Ellison's truth: infinite possibilities, just around the corner.

Onto this scene steps *The Book of Rhythms.* No doubt Hughes viewed it somewhat as his hypercautious publishers had done—as a way to reach the children's market without ruffling the feathers of conservative parents, teachers, librarians, politicians, and others on guard against "communists." Accordingly, Hughes's central motive was to do what he had done at the Lab School and the College Language Association conference, only this time in a children's book format: he would present a lesson associating rhythms in poetry with rhythms in nature and in the expressions of cultures throughout the world (with an emphasis on the United States).

Rhythms was one of the five books that Hughes wrote for Franklin Watts's series of First Books for children. After *The First Book of Negroes* (1952) and *The First Book of Rhythms* (1954), Hughes also published *The First Book of Jazz* (1955),

The First Book of the West Indies (1956), and *The First Book of Africa* (1960). The jacket of the original *Book of Rhythms* announced the purpose of the series in bold type: "When boys and girls FIRST start asking why?...what?...and how? FIRST BOOKS are the first books to read on any subject." The series presented First Books on bugs, stones, and prehistoric animals; on electricity, space travel, and water; dolls, magic, and photography. Under the category "People Around the World" were listed First Books on Hawaii, Eskimos, Israel, Mexico, and Negroes. (Curiously, *The First Book of Negroes* was the only "racial," as opposed to national or regional, title in the series.) *The Book of Rhythms* fell into the series' list of "New Worlds to Know and Enjoy," a category that included *Mythology, Words,* and Hughes's *Jazz.* In this music-dance-literature subsection of the First Books catalog, *Rhythms* took its place as "a lyric interpretation of life and its varied beats."

Since *Rhythms* was not overtly a controversial book, it suffered none of the editorial excisions of *The First Book of Negroes, Famous American Negroes,* or *The First Book of Jazz.* (Apparently, the only cut made by the publisher was in an early section on the moon, where Hughes, tongue in cheek, warned that on nights with a full moon, certain people's brains seemed to get pulled out of whack!)

In the face of the apparent shunning of "radical" politics here, what is most fascinating about this fine little volume is its spirited play of issues and words that suggest this period of potentiality and incipient political change in black America. There is a sense in which Hughes did indeed manage to offer, in language nearly as subtle as a Miles Davis solo, a profoundly political "interpretation of life and its various beats." Operating, so to speak, *invisibly,* Hughes delivered in *The Book of Rhythms* a complex statement that belongs, in its way, to the era of Rosa Parks, Martin Luther King, Jr., and the knockout swings of Sugar Ray Robinson and Willie Mays.

To understand this perspective on *The Book of Rhythms,* recall the stereotype of Africans (New World or Old World) as people of "natural rhythm." Without ever referring to this racist myth, Hughes's text quietly undercuts it by defining rhythm as originating not in "race" but in the patterned shapes and movements of nature (the stars, leaves, winds, and rivers, for example), in the pulses of the human body, and in learned cultural practices. Thus Hughes contradicts a very old tenet of antiblack racism and, implicitly, the very idea of race as a category of meaning.

Addressing his young readers, whatever their color or nationality, Hughes says: "Your rhythm on this earth began first with the beat of your heart.... The rhythm of life is the beat of the heart." Invoking one anthropological theory of the origins of artistic expression (in music and dance, often referred to as antecedents of poetry and drama), Hughes goes on to say that "thousands of years ago man transformed the rhythm of the heartbeat into a drumbeat, and the rhythm of music began. They made a slow steady drumbeat to walk to or march to, a faster beat to sing to, and a changing beat to dance to." If rhythm is ever "natural," then it is universally so: although playing a percussive instrument and dancing to the beat are learned cultural specialties, having some sense of rhythm is as natural to the *human* race as the heartbeat and the perception of a surging breeze.

Just as Hughes takes for granted that rhythm is an aspect of the human race, he assumes that personal rhythms suggest not ethnic or national but *individual culture.* Make a set of circles inside larger circles, Hughes tells his readers. "See how these circles almost seem to move, for you have left something of your own movement there, and your own feeling of place and of roundness. Your circles are not quite like the circles of anyone else in the world, because you are not like anyone else.... Your circles and rhythms are yours alone." During this period of violent denial of black humanity and individuality, Hughes makes his erasure of racial categories a purposeful silence that suggests a worldview permanently at odds with systems of race division and classification. Rhythm is universal and yet it declares the independence of individuality: *"Your circles and rhythms are yours alone."*

Hughes's text also offers a very hopeful vision of the world as a single community. Humans are the animals that can perceive rhythms and then reenact them in various forms of art. In this way, according to *The Book of Rhythms,* Paul Manship's statue of Prometheus in New York's Rockefeller Plaza is connected to Picasso's stylized bull, El Greco's created man, and the art of Africa, where "artists a thousand years ago made beautiful masks with rhythmical lines." And elsewhere in the book: "Perhaps the curve of a waterfall or the arching stripes of the rainbow suggested the rhythms for...the arch of Tamerlane's tomb at Samarkand, the arch of a bridge in ancient China, or the Moorish arches at Granada."

Likewise, the vision of the United States that is projected here is boldly cosmopolitan and many-hued in its cultural shapes and components: its rhythms. In the section on "How Rhythms Take Shape," Hughes artfully inserts his pluralistic vision into a discussion of clothing styles, music, and architecture:

> Fashions in clothes and styles of decoration and design are everywhere influenced by other people and places around the world. That design on your father's tie may be from Persia. Your uncle's jacket is a Scotch plaid. The lines of your mother's dress are French. Her shawl is Spanish in design.
>
> The music on your radio now is Cuban, its drums are the bongos of Africa, but the orchestra playing it is American. Rhythms go around the world, adopted and molded by other countries, mixing with other rhythms, and creating new rhythms as they travel....
>
> The Quonset hut of the American army is in shape not unlike an Eskimo igloo. And the East Indians are building skyscrapers in Bombay. We love Irish folk songs. The Irish love American jazz. Sometimes American jazz and Irish melodies combine to make a single song.

This view of American culture, and indeed of world culture, as a beautifully rhythmic patchwork quilt is supported by this volume's assertion of a United States still resolving issues dating from the American Revolution and the Civil War. In his pages on poetry, Hughes encloses some "rolling lines" from Walt Whitman's "When Lilacs Last in the Dooryard Bloom'd," that impassioned elegy to the memory of Abraham Lincoln, symbol of the struggle for American unity and black freedom. Lest the point be too subtly traced, Hughes follows up the Whitman passage with another invocation of the Civil War in the form of classic lines from Lincoln's Gettysburg Address describing America's genesis as "a new nation, conceived in liberty, and dedicated to the proposition that all men are created equal." In 1954, these familiar words had deep resonance (as well as an echo of irony) in the ears of black Americans.

Other themes enrich *The Book of Rhythms:* the implication that not only art but human consciousness itself derives from the human body (that the metaphysical is inextricable from the physical); the connections between modern design in buildings, cars, planes, and furniture and the brilliant simplicity of rhythms in nature.

There are moments of disappointment, too: Hughes implies that art, like flowers in the sun, "naturally" points upward (thus the skyscraper, like the ballet dancer, reaches up and points upward). But in fact this view is culturally specific and suggests a problem when one counts too much on the romantic equation of perceived "universal" patterns in nature to patterns in human culture. For instance, the drummers and dancers in many African cultures stress not the upness but the downness of much cultural expression. Senegalese dancers bend arms and knees as they gesture, like their drummers, to the dark earth. In the down-home spaces and the Harlems of America, too, what is most valued is not only the capacity in art to get further and further up, but, as the expression goes, to get down.

Still, *The Book of Rhythms* is rich with meanings that have been made more evident with the passage of time. Saved, as in time-release capsules, these subtleties give the book a quiet, steady power. Obviously, Hughes sent forth this slender book to reach a paying audience and to make the rent. (One of his poems declares: "I wish the rent / Was heaven sent.") But *The Book of Rhythms* also was a dedicated teacher's master class on the poetry of rhythm through the world. And it helped to get Americans ready for integration on a social and political level that he already detected on a cultural level.

Rhythm, as Hughes defined it, linked us all to the earth and the stars and "with all the plants and animals and people in the world." Prophetically, he saw this grand physical/metaphysical power in the hands of young readers on the brink of the Civil Rights Revolution. Your rhythm, he told them, "is related to the rhythms of the earth as it moves around the sun, and to the moon...[and to] those vaster rhythms of time and space and wonder beyond the reach of eye and mind." In your hands, he implied, is the way to make a world that is more unified and free, a new world pulsing with possibilities for all.

Langston Hughes (1902–1967) was one of the most versatile writers of the artistic movement known as the Harlem Renaissance. Though known primarily as a poet, Hughes also wrote plays, essays, novels, and short stories. He was a well-regarded guest teacher at the University of Chicago Lab School in the 1940s and wrote many books for children, including *The First Book of Africa, The First Book of Jazz, Popo and Fifina* (with Arna Bontemps), *Black Misery*, and *The Sweet and Sour Animal Book.* With Bontemps, he also compiled the anthologies *Poetry of the Negro* and *The Book of Negro Folklore.*

Wynton Marsalis is a highly acclaimed jazz and classical trumpeter. He grew up in Kenner, Louisiana, and was influenced by a wide range of musical styles—including traditional New Orleans jazz, bebop, rhythm and blues, and classical. In 1979 he entered the Juilliard School in New York City. In 1980 Marsalis's professional music career was launched when he became the regular trumpet player with Art Blakey's Jazz Messengers. He has since gone on to record many notable albums, both classical and jazz. Marsalis also serves as Artistic Director for Jazz at Lincoln Center. In that capacity, he hosts several "Jazz for Young People" concerts each year.

Robert G. O'Meally is Zora Neale Hurston Professor of American Literature at Columbia University and previously taught English and Afro-American studies at Wesleyan University and Barnard College. He is the author of *The Craft of Ralph Ellison* and *Lady Day: Many Faces of the Lady* and editor of *Tales of the Congaree* by E. C. Adams and *New Essays on "Invisible Man."* Professor O'Meally is coeditor of *History and Memory in African American Culture* and *Critical Essays on Sterling A. Brown.*

Matt Wawiorka is a free-lance artist in Kenosha, Wisconsin. Some of the other books that he has illustrated are *Five Minute Miracles, A Teen's Guide to Going Vegetarian, Something Spicy,* and *Something Sweet.* He also creates illustrations for magazines, annual reports, and advertising agencies.

THE IONA AND PETER OPIE LIBRARY OF CHILDREN'S LITERATURE

The Opie Library brings to a new generation an exceptional selection of children's literature, ranging from facsimiles and new editions of classic works to lost or forgotten treasures—some never before published—by eminent authors and illustrators. The series honors Iona and Peter Opie, the distinguished scholars and collectors of children's literature, continuing their lifelong mission to seek out and preserve the very best books for children.

ROBERT G. O'MEALLY
GENERAL EDITOR